SURVEYING MATHEMATICS MADE SIMPLE

An original Book by

Jim Crume P.L.S., M.S., CFedS

Co-Authors
Cindy Crume
Bridget Crume
Troy Ray R.L.S.
Mark Sandwick P.L.S.
Mark Lull

I0470479

KINDLE - PRINTED EDITIONS

*****

PUBLISHED BY:

Jim Crume P.L.S., M.S., CFedS

Crume's Transformation

First publication: October, 2015

Cover photo: S.R. 202
Mesa, Arizona

TERMS AND CONDITIONS

TABLE OF CONTENTS

INTRODUCTION

Straight forward Step-by-Step instructions.

This book is just one part in a series of digital and paperback books on Surveying Mathematics Made Simple. The subject matter in this book will utilize the methods and formulas that are covered in the books that precede it. If you have not read the preceding books, you are encouraged to review a copy before proceeding forward with this book.

For a list of books in this series, please visit:

http://www.cc4w.net/ebooks.html

Prerequisites for this book:

A basic knowledge of geometry, algebra, trigonometry, COGO, CADD and Spreadsheet programs are required for the explanations shown in this book.

The following books of this series are recommended to complete some of the step by step processes in this book:

Bearings and Azimuths - Book 1
Create Rectangular Coordinates - Book 2
Inverse Between Rectangular Coordinates - Book 3
Circular Curves - Book 4
Spiral Curves - Book 6
The Myth About Spiral Curve Offsets - Book 7
Coordinate Transformation - Book 9
Highway Centerline (Retracement) - Book 14

These books contain formulas, step by step solutions and examples.

DEFINITIONS

Best Fit: Is a mathematical function to fit a line, lines and/ or curves through a set of points representing existing monuments that minimizes the *residuals (see definition)* when comparing record offset distances to measured distances and simultaneously minimizing the stationing residuals between record and measured.

CAD or CADD: Computer-aided design (CAD) and drafting (CADD) is the use of computer systems to assist in the graphical creation, modification, analysis, or optimization of a design.

COGO: A Coordinate Geometry program designed to calculate rectangular coordinates, simple and spiral curves, parcel boundaries and areas, centerline alignments, station and offset reports, and other surveying related computations.

Controlling Centerline: A record centerline alignment from which the right of way corridor left and right of said centerline is measured.

Coordinates (Final): Final coordinates are points that represent the true position of a point whose values are fixed either by a found monument or calculated position.

Coordinates (Search): Search coordinates are points that are temporary, whose positions have been generated by approximate methods in order to navigate to an area to look for existing monumentation.

Edge of Pavement or Striping: The visible edge or striping, left and right, of a paved or dirt road that are of a specific distance from the centerline of the road that can be used to mathematically determine the centerline of the road.

Fences (Existing): Are fences that purport to represent the right of way corridor and are of a specific distance from the centerline of the road that can be used to mathematically determine the centerline of the road.

Monuments (Existing): Found monuments that are located by survey measurements that are referenced to a specified horizontal datum that purport to define or have a relationship with the record centerline and/or right of way corridor.

Right of Way Corridor: Is the limits of the right of way boundary on the Left and Right side of the controlling centerline for a highway or road.

Residuals: Is the difference between the calculated position and actual position of an existing monument both in distance, offset and stationing from a centerline (record or calculated).

 Throughout this book, tips will be given to help explain or give directions on the subject matter.

 This Icon is for CADD tips or directions that will be given throughout this book.

Reference to SimPro-GC Suite and/or it's modules.

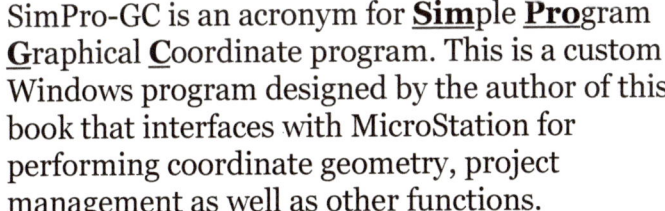

SimPro-GC is an acronym for **Sim**ple **Pro**gram **G**raphical **C**oordinate program. This is a custom Windows program designed by the author of this book that interfaces with MicroStation for performing coordinate geometry, project management as well as other functions.

This program is not needed for examples shown in this book. Any coordinate geometry program will work as long as there is a process to get the points that are calculated into your CADD platform.

Typical Numbers (Mac) spreadsheet formula reference:

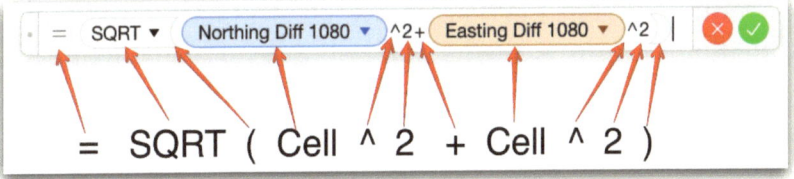

Typical Excel (Windows) spreadsheet formula reference:

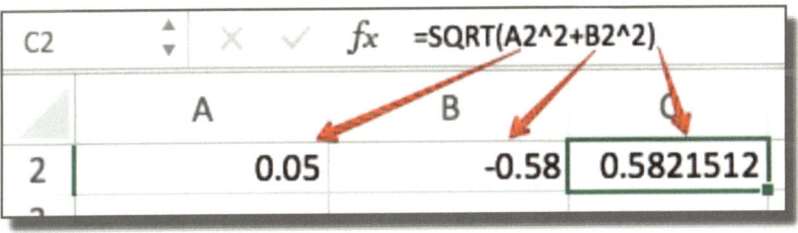

Note: Consult the user's manual for the available functions in Numbers and Excel that correspond with the formulas shown in this book.

CRUME'S TRANSFORMATON

The term "Best Fit" can mean almost anything depending upon who you ask. The question becomes, how do you quantify a "Best Fit" or in other words how do you prove that you have determined a "Best Fit"?

The best possible solution is to show mathematically that you have determined a "Best Fit".

I have spent many years researching, talking to other professionals and applying "Best Fit" theories and procedures to determine controlling centerlines.

I have and do use Least Squares adjustment which is a cool tool. It doesn't work for every application such as a centerline alignment with multiple tangents and curves with offset fixed points. Linear Regression, which is a Least Squares adjustment for lines, works well but only for a single line at a time. Least Squares adjustment can be complicated especially when you throw in weighted points which is subjective by the operator and is not always repeatable by others. Algebraic or Geometric Curve Fitting (using Least Squares) will not work with horizontal and spiral curve alignments and still maintain the record geometry such as the Radius length or other curve constraints.

Least Squares is not easily adapted to a spreadsheet program nor is it understood by all surveyors, engineers and GIS professionals.

I wanted a procedure that would work with right of way corridors, that would be easy to use and could be shown mathematically using simple statistics and adaptable to a spreadsheet program like Numbers or Excel to come up with a "Best Fit" solution.

I could have called the solution "The Statistical Analysis of Determining a Best Fit solution for Right of Way Corridors". Instead, I chose a much simpler name of "Crume's Transformation". I could have also written the process in boring mathematical terms and symbols that would put most people to sleep. I chose to keep with the theme of my books **"Surveying Mathematics Made Simple"** and explain the process using spreadsheets and simple terms.

Crume's Transformation has two methods, **Fixed Points** and **Random Points**. Fixed Points is for best fitting found monumentation with record geometry. Random Points is to best fit record geometry to fence lines, edge of pavement, lane stripes and centerline stripes. Both methods will be presented here.

There is a simple video that demonstrates the Fixed Points Transformation locate at:

<u>https://youtu.be/AGFQvVhe4Mg</u>

You need to have knowledge of Rectangular Coordinates and access to a Coordinate Geometry program to generate coordinates to apply this transformation.

You also need knowledge of how to utilize spreadsheets and entering formulas.

This transformation and the process to perform a highway retracement survey is covered in great detail in Book 14 "Highway Centerlines (Retracement)".

FIXED POINTS METHOD

Identify the found monumentation that corresponds to the record geometry as shown on the highway right of way plans, results of survey or other documentation. Generate (search or temporary) rectangular coordinates for the record geometry and offset points.

Create a table in Numbers (Mac) or Excel (Windows) or other spreadsheet program named "Found and Search Positions".

Place the Found Monumentation rectangular coordinates in two columns of Northing and Easting.

Place the corresponding (search or temporary) rectangular coordinates based upon record geometry in two columns of Northing and Easting.

Create two new columns labeled Northing Diff and Easting Diff. Place formulas in these cells that subtract the Search Northing from the Found Northing. Repeat for the Easting coordinates. Copy the formulas in these two cells and paste them into the cells for each of the remaining points in the data set.

(Pt's 1080 to 1215 are the found monument numbers.)

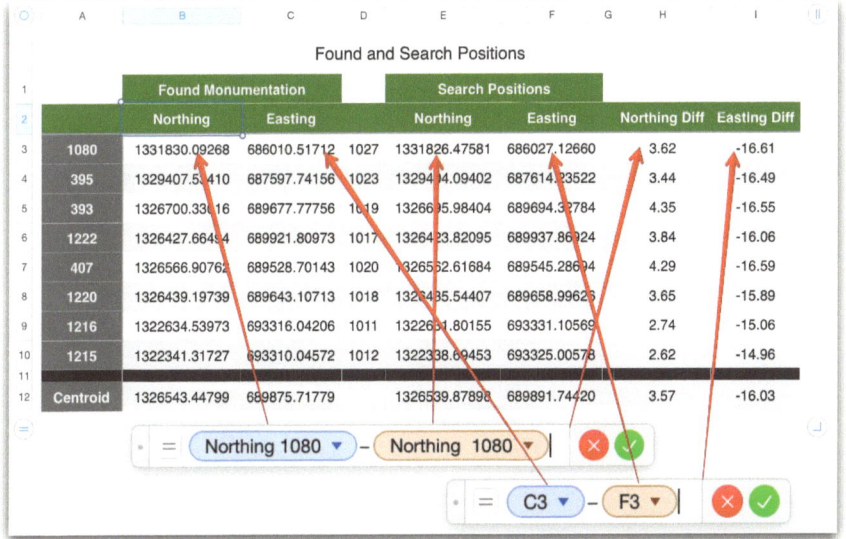

Northing Diff = Found Northing - Search Northing
Easting Diff = Found Easting - Search Easting

Repeat for each point in the data set.

Add a row labeled Centroid. In this row calculate the Mean (Average) of each of the Northing and Easting coordinate columns for the Found and Search coordinates.

The Centroid is the mathematical center of the coordinate system. The Centroid is the origin point for which the rotation and northerly/easterly shift will be applied later on in this process.

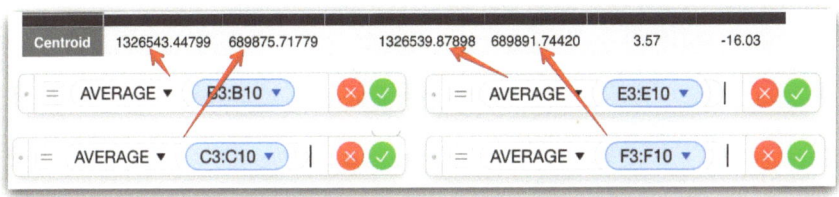

Northing Centroid = $(N_1 + N_2 + ... N_n) / n$
Easting Centroid = $(E_1 + E_2 + ... E_n) / n$

Where (n) is the number of points

Next calculate the difference between the Centroid for the found and search coordinates.

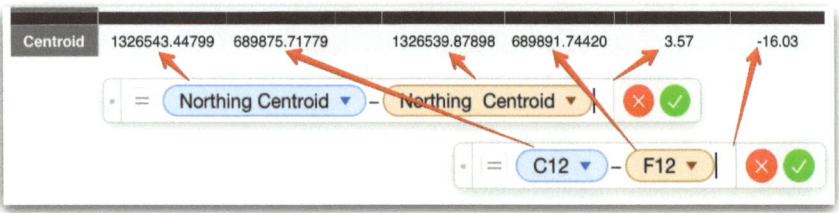

Centroid Northing Diff (CND) = Found Centroid Northing - Search Centroid Northing
Centroid Easting Diff (CED) = Found Centroid Easting - Search Centroid Easting

Next create a new table for synchronizing the Centroids named "Synchronized Centroids".

Copy the Found Monumentation columns to this table. Create two new columns for the Translated Northing and Easting values.

In these cells, add the Search Northing Coordinate plus the Northing Difference for the Centroid. Repeat this process for the Search Easting Coordinate.

Crume's Transformation

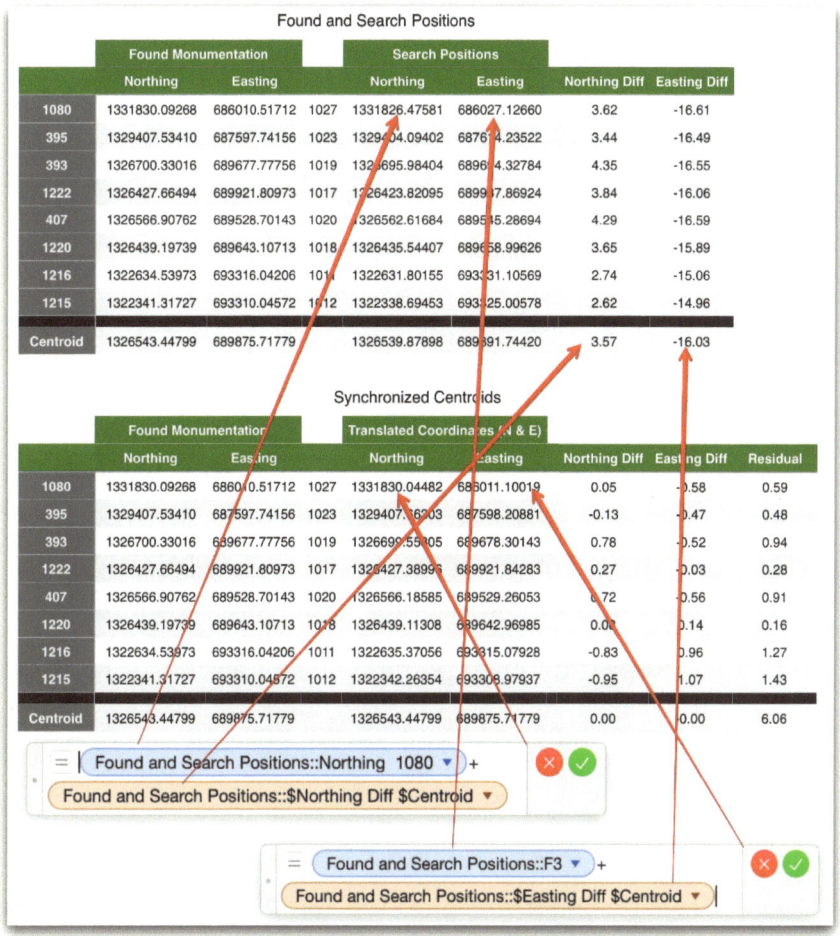

Found and Search Positions

	Found Monumentation			Search Positions		Northing Diff	Easting Diff
	Northing	Easting		Northing	Easting		
1080	1331830.09268	686010.51712	1027	1331826.47581	686027.12660	3.62	-16.61
395	1329407.53410	687597.74156	1023	1329404.09402	68764.23522	3.44	-16.49
393	1326700.33016	689677.77756	1019	1326695.98404	68964.32784	4.35	-16.55
1222	1326427.66494	689921.80973	1017	1326423.82095	68997.86924	3.84	-16.06
407	1326566.90762	689528.70143	1020	1326562.61684	68945.28694	4.29	-16.59
1220	1326439.19739	689643.10713	1018	1326435.54407	68968.99626	3.65	-15.89
1216	1322634.53973	693316.04206	1011	1322631.80155	69331.10569	2.74	-15.06
1215	1322341.31727	693310.04572	1012	1322338.69453	69325.00578	2.62	-14.96
Centroid	1326543.44799	689875.71779		1326539.87898	68991.74420	3.57	-16.03

Synchronized Centroids

	Found Monumentation			Translated Coordinates (N & E)		Northing Diff	Easting Diff	Residual
	Northing	Easting		Northing	Easting			
1080	1331830.09268	686010.51712	1027	1331830.04482	686011.10019	0.05	-0.58	0.59
395	1329407.53410	687597.74156	1023	1329407.6203	687598.20881	-0.13	-0.47	0.48
393	1326700.33016	689677.77756	1019	1326699.55105	689678.30143	0.78	-0.52	0.94
1222	1326427.66494	689921.80973	1017	1326427.38996	689921.84283	0.27	-0.03	0.28
407	1326566.90762	689528.70143	1020	1326566.18585	689529.26053	0.72	-0.56	0.91
1220	1326439.19739	689643.10713	1018	1326439.11308	689642.96985	0.00	0.14	0.16
1216	1322634.53973	693316.04206	1011	1322635.37056	693315.07928	-0.83	0.96	1.27
1215	1322341.31727	693310.04572	1012	1322342.26354	693308.97937	-0.95	1.07	1.43
Centroid	1326543.44799	689875.71779		1326543.44799	689875.71779	0.00	0.00	6.06

```
= ( Found and Search Positions::Northing  1080 ▼ ) +
  Found and Search Positions::$Northing Diff $Centroid ▼
```

```
= ( Found and Search Positions::F3 ▼ ) +
  Found and Search Positions::$Easting Diff $Centroid ▼
```

Translated Northing = Search Northing + CND
Translated Easting = Search Easting + CED

Repeat for each point in the data set.

Next calculate the difference for the Northing and Easting of the Found Monumentation and Translated Coordinates.

Crume's Transformation

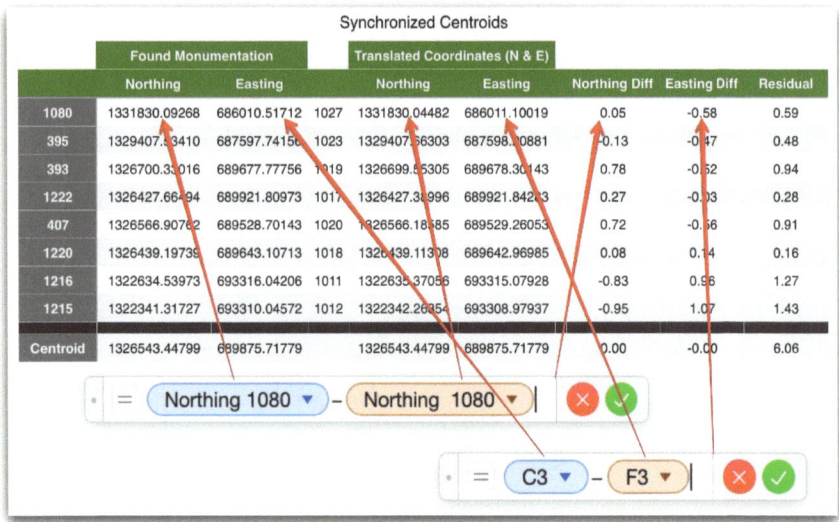

Northing Diff = Found Northing - Translated Northing
Easting Diff = Found Easting - Translated Easting

Repeat for each point in the data set.

Create a new column for the Residual.

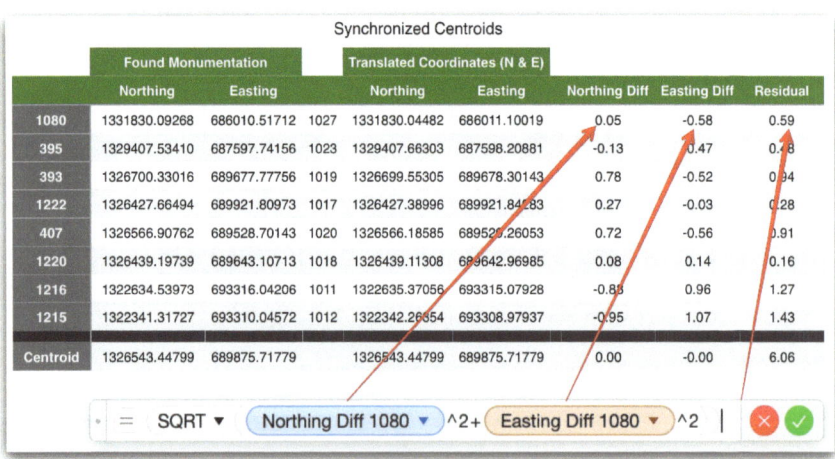

Residual = SQRT(Northing Diff^2 + Easting Diff^2)

Repeat this for each point in the data set.

The Residual is the distance from the Found Monumentation to the Translated Coordinate position using the Pythagorean Theorem.

If the monumentation was set and measured perfectly, then the coordinates for the found monuments and translated coordinates would be the same and would have a 0.00 residual.

In the real world, monuments are never set perfectly and they cannot be measured perfectly due to the limitations of the survey equipment, systematic errors and random errors. Therefore a mathematical process needs to be implemented that will account for these limitations and errors.

 **Note**: The difference between the Centroid of the Found Monumentation and the Translated Coordinate position should be 0.00 indicating that the Centroids are synchronized.

Create a new table named "Centroid to Translated Point". In this table calculate the "Latitude" and "Departure" between the Translated Centroid and Translated Coordinates and place them in separate columns.

Crume's Transformation

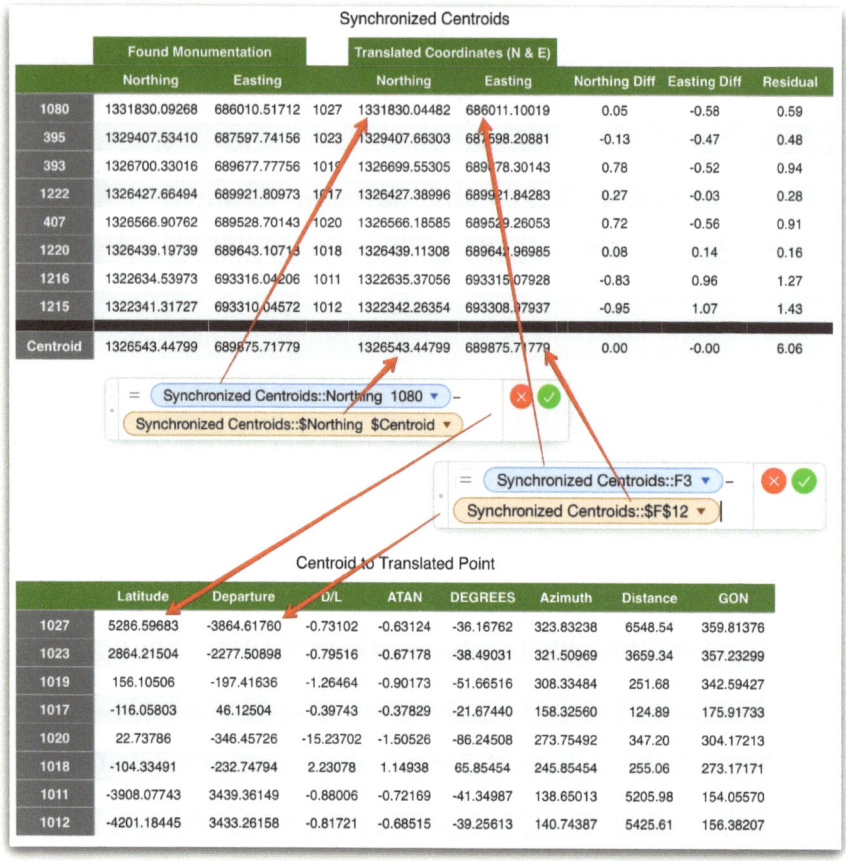

	Found Monumentation			Translated Coordinates (N & E)				
	Northing	Easting		Northing	Easting	Northing Diff	Easting Diff	Residual
1080	1331830.09268	686010.51712	1027	1331830.04482	686011.10019	0.05	-0.58	0.59
395	1329407.53410	687597.74156	1023	1329407.66303	687598.20881	-0.13	-0.47	0.48
393	1326700.33016	689677.77756	1019	1326699.55305	689678.30143	0.78	-0.52	0.94
1222	1326427.66494	689921.80973	1017	1326427.38996	689921.84283	0.27	-0.03	0.28
407	1326566.90762	689528.70143	1020	1326566.18585	689529.26053	0.72	-0.56	0.91
1220	1326439.19739	689643.10713	1018	1326439.11308	689642.96985	0.08	0.14	0.16
1216	1322634.53973	693316.04206	1011	1322635.37056	693315.07928	-0.83	0.96	1.27
1215	1322341.31727	693310.04572	1012	1322342.26354	693308.97937	-0.95	1.07	1.43
Centroid	1326543.44799	689875.71779		1326543.44799	689875.71779	0.00	-0.00	6.06

= (Synchronized Centroids::Northing 1080 ▾) –
(Synchronized Centroids::$Northing $Centroid ▾)

= (Synchronized Centroids::F3 ▾) –
(Synchronized Centroids::$F$12 ▾)

Centroid to Translated Point

	Latitude	Departure	D/L	ATAN	DEGREES	Azimuth	Distance	GON
1027	5286.59683	-3864.61760	-0.73102	-0.63124	-36.16762	323.83238	6548.54	359.81376
1023	2864.21504	-2277.50898	-0.79516	-0.67178	-38.49031	321.50969	3659.34	357.23299
1019	156.10506	-197.41636	-1.26464	-0.90173	-51.66516	308.33484	251.68	342.59427
1017	-116.05803	46.12504	-0.39743	-0.37829	-21.67440	158.32560	124.89	175.91733
1020	22.73786	-346.45726	-15.23702	-1.50526	-86.24508	273.75492	347.20	304.17213
1018	-104.33491	-232.74794	2.23078	1.14938	65.85454	245.85454	255.06	273.17171
1011	-3908.07743	3439.36149	-0.88006	-0.72169	-41.34987	138.65013	5205.98	154.05570
1012	-4201.18445	3433.26158	-0.81721	-0.68515	-39.25613	140.74387	5425.61	156.38207

Latitude = (Translated Northing - Centroid Northing)
Departure = (Translated Easting - Centroid Easting)

Repeat this for each point in the data set.

(Pt's 1027 to 1012 are the search point numbers that correspond to the found monument location. You can assign any value you want for the row ID number.)

Create a new column "D/L" then divide the Departure by the Latitude to calculate the trigonometric relationship for each point in the data set.

Crume's Transformation

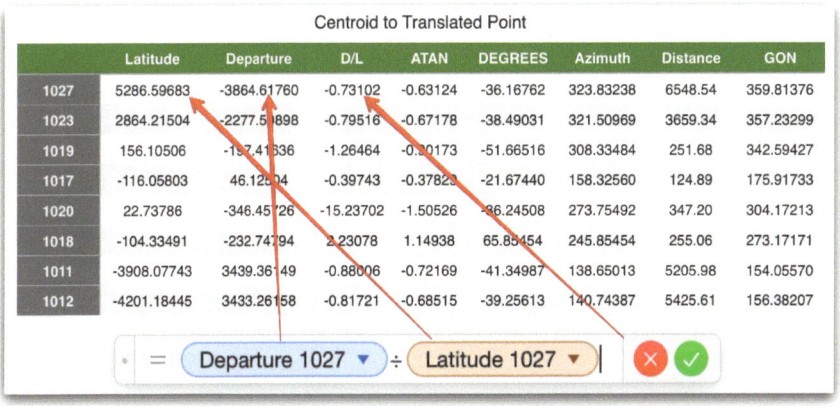

$$D/L = \text{Departure} / \text{Latitude}$$

Repeat for each point in the data set.

Create a new column "ATAN" then take the ArcTan of the "D/L" value to calculate the angle in **Radians** for each point in the data set.

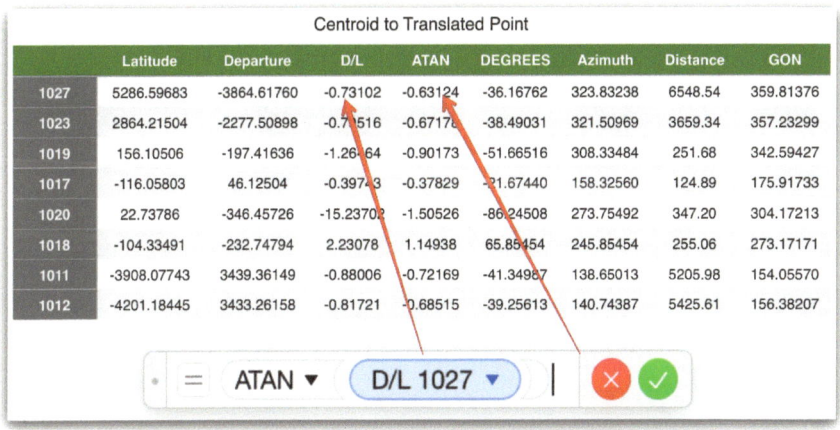

$$\text{ATAN} = \text{ArcTan}(\, D/L \,)$$

Repeat for the each point in the data set.

DEGREES

Create a new column "DEGREES" then convert Radians to Degrees for each point in the data set. (See below for GONS conversion)

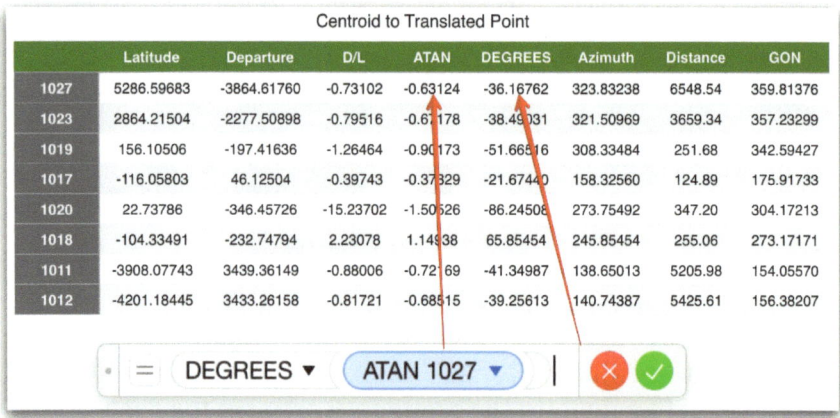

DEGREES = DEGREES (ATAN)

Repeat for each point in the data set.

Create a new column "Azimuth" then convert Degrees to Azimuth for each point in the data set.

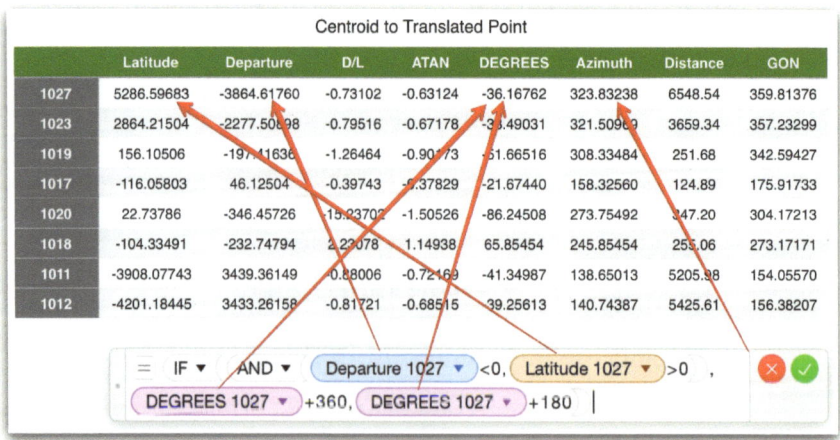

The formula in this column has two different solutions that are controlled by the "IF" and "AND" statements.

If Departure is < 0 and Latitude > 0, then add 360 to DEGREES, or else add 180 to DEGREES.

You will need to modify this formula as needed depending upon the quadrant you are working in. (i.e. for the NE quadrant the Azimuth will equal the Bearing).

See Bearings and Azimuths - Book 1 for more information to convert from Bearings to Azimuths.

Create a new column "Distance" then calculate the Distance using the Pythagorean Theorem.

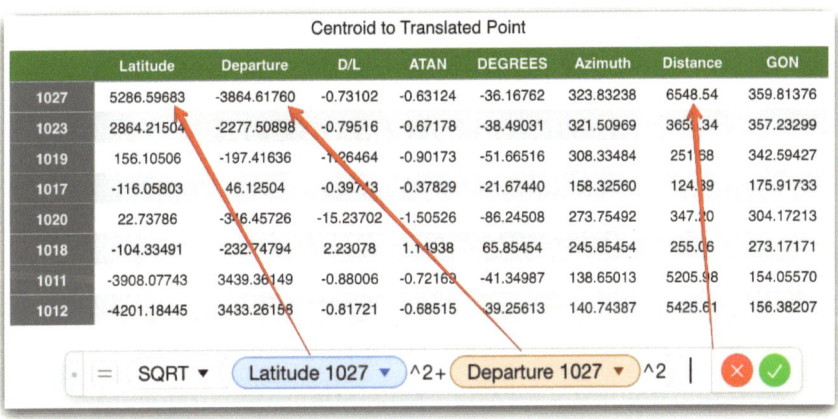

	Latitude	Departure	D/L	ATAN	DEGREES	Azimuth	Distance	GON
1027	5286.59683	-3864.61760	-0.73102	-0.63124	-36.16762	323.83238	6548.54	359.81376
1023	2864.21504	-2277.50898	-0.79516	-0.67178	-38.49031	321.50969	3659.34	357.23299
1019	156.10506	-197.41636	-1.26464	-0.90173	-51.66516	308.33484	251.68	342.59427
1017	-116.05803	46.12504	-0.39743	-0.37829	-21.67440	158.32560	124.89	175.91733
1020	22.73786	-346.45726	-15.23702	-1.50526	-86.24508	273.75492	347.20	304.17213
1018	-104.33491	-232.74794	2.23078	1.14938	65.85454	245.85454	255.06	273.17171
1011	-3908.07743	3439.36149	-0.88006	-0.72169	-41.34987	138.65013	5205.98	154.05570
1012	-4201.18445	3433.26156	-0.81721	-0.68515	39.25613	140.74387	5425.61	156.38207

= SQRT ▼ (Latitude 1027 ▼)^2+ (Departure 1027 ▼)^2 |

Distance = SQRT(Latitude^2 + Departure^2)

Repeat this for each point in the data set.

GONS

Create a new column "GON" then calculate the GON value for each point in the data set.

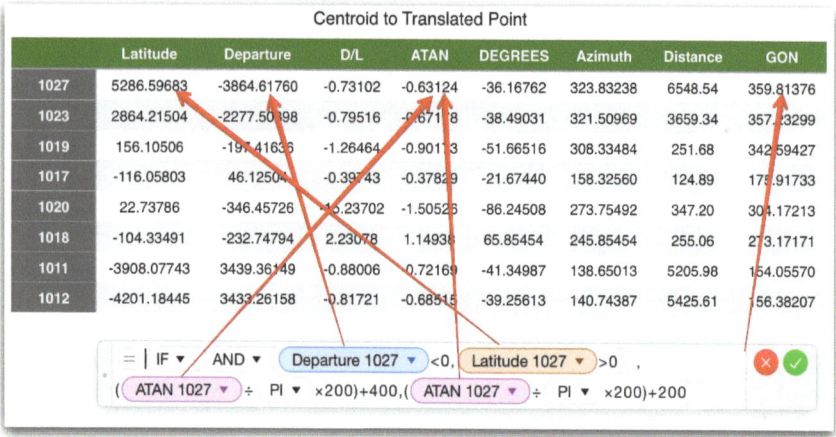

Centroid to Translated Point

	Latitude	Departure	D/L	ATAN	DEGREES	Azimuth	Distance	GON
1027	5286.59683	-3864.61760	-0.73102	-0.63124	-36.16762	323.83238	6548.54	359.81376
1023	2864.21504	-2277.50198	-0.79516	-0.6718	-38.49031	321.50969	3659.34	357.13299
1019	156.10506	-197.41636	-1.26464	-0.90113	-51.66516	308.33484	251.68	342.59427
1017	-116.05803	46.12504	-0.39743	-0.37829	-21.67440	158.32560	124.89	175.91733
1020	22.73786	-346.45726	-15.23702	-1.50526	-86.24508	273.75492	347.20	304.17213
1018	-104.33491	-232.74794	2.23078	1.14936	65.85454	245.85454	255.06	273.17171
1011	-3908.07743	3439.36149	-0.88006	0.72169	-41.34987	138.65013	5205.98	154.05570
1012	-4201.18445	3433.26158	-0.81721	-0.68515	-39.25613	140.74387	5425.61	156.38207

= | IF ▼   AND ▼   (Departure 1027 ▼) <0,   (Latitude 1027 ▼) >0 ,

(ATAN 1027 ▼) ÷ PI ▼ ×200)+400,((ATAN 1027 ▼) ÷ PI ▼ ×200)+200

If Departure is < 0 and Latitude > 0, convert ATAN column (in Radians) to GON, then add 400 to get GONS or else add 200 to get GONS.

You will to need to modify this formula as needed depending upon the quadrant you are working in.

BRINGING IT ALL TOGETHER

Create a table named "3 Parameter - Rotation and Translation of calculated positions".

Copy the Found Monumentation Northing, Easting and Centroid cells.

Create two columns for the Translated Coordinates for the Northing and Easting.

Create a row for the Rotation value.

Create a row for the Northing and Easting shift.

DEGREES

Add the following formulas that this will reference the other tables previously created to these cells:

(See below for the GON solution)

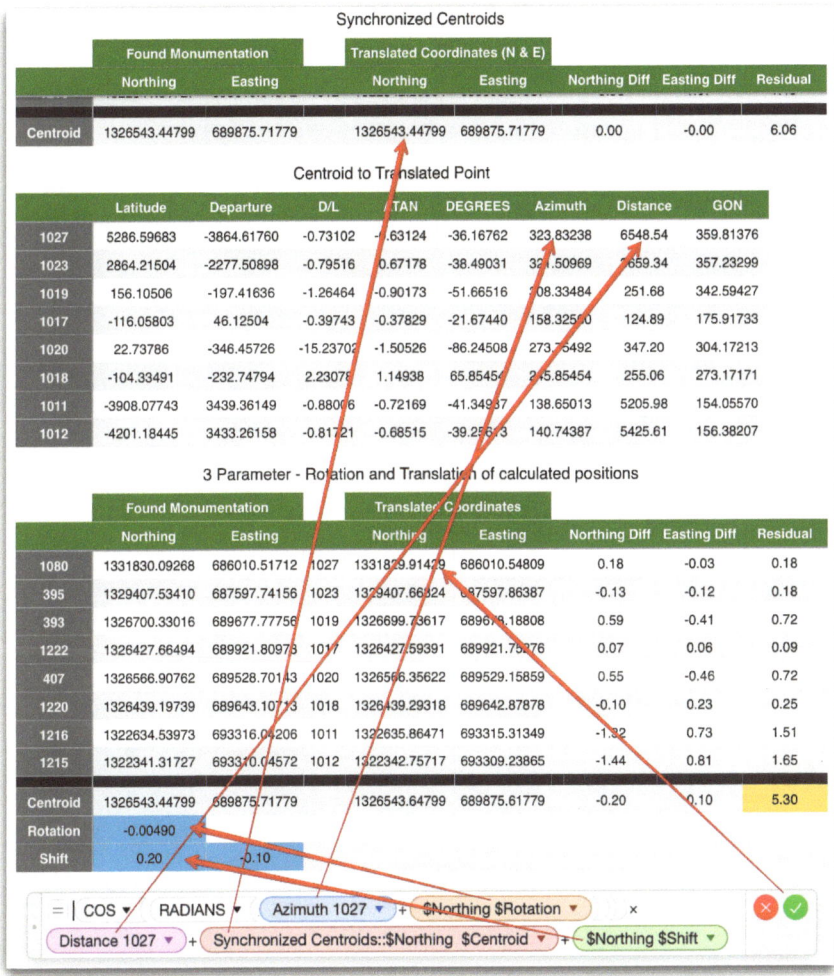

Synchronized Centroids

	Found Monumentation			Translated Coordinates (N & E)				
	Northing	Easting		Northing	Easting	Northing Diff	Easting Diff	Residual
Centroid	1326543.44799	689875.71779		1326543.44799	689875.71779	0.00	-0.00	6.06

Centroid to Translated Point

	Latitude	Departure	D/L	TAN	DEGREES	Azimuth	Distance	GON
1027	5286.59683	-3864.61760	-0.73102	-0.63124	-36.16762	323.83238	6548.54	359.81376
1023	2864.21504	-2277.50898	-0.79516	0.67178	-38.49031	321.50969	3659.34	357.23299
1019	156.10506	-197.41636	-1.26464	-0.90173	-51.66516	308.33484	251.68	342.59427
1017	-116.05803	46.12504	-0.39743	-0.37829	-21.67440	158.32550	124.89	175.91733
1020	22.73786	-346.45726	-15.23702	-1.50526	-86.24508	273.75492	347.20	304.17213
1018	-104.33491	-232.74794	2.23078	1.14938	65.85454	245.85454	255.06	273.17171
1011	-3908.07743	3439.36149	-0.88006	-0.72169	-41.34937	138.65013	5205.98	154.05570
1012	-4201.18445	3433.26158	-0.81721	-0.68515	-39.25663	140.74387	5425.61	156.38207

3 Parameter - Rotation and Translation of calculated positions

	Found Monumentation			Translated Coordinates				
	Northing	Easting		Northing	Easting	Northing Diff	Easting Diff	Residual
1080	1331830.09268	686010.51712	1027	1331829.91429	686010.54809	0.18	-0.03	0.18
395	1329407.53410	687597.74156	1023	1329407.66324	687597.86387	-0.13	-0.12	0.18
393	1326700.33016	689677.77756	1019	1326699.73617	689678.18808	0.59	-0.41	0.72
1222	1326427.66494	689921.80978	1017	1326427.59391	689921.75276	0.07	0.06	0.09
407	1326566.90762	689528.70143	1020	1326566.35622	689529.15859	0.55	-0.46	0.72
1220	1326439.19739	689643.10713	1018	1326439.29318	689642.87878	-0.10	0.23	0.25
1216	1322634.53973	693316.04206	1011	1322635.86471	693315.31349	-1.32	0.73	1.51
1215	1322341.31727	693310.04572	1012	1322342.75717	693309.23865	-1.44	0.81	1.65
Centroid	1326543.44799	689875.71779		1326543.64799	689875.61779	-0.20	0.10	5.30
Rotation	-0.00490							
Shift	0.20	-0.10						

= | COS ▾ | RADIANS ▾ | (Azimuth 1027 ▾) + $Northing $Rotation ▾) × ❌ ✅
(Distance 1027 ▾) + (Synchronized Centroids::$Northing $Centroid ▾) + ($Northing $Shift ▾)

Translated Northing = Cos(toRadians(Azimuth + Rotation)) x Distance + Centroid Northing + Northing Shift

Repeat for each point in the data set.

Synchronized Centroids

	Found Monumentation			Translated Coordinates (N & E)				
	Northing	Easting		Northing	Easting	Northing Diff	Easting Diff	Residual
1215	1322341.01727	093310.04372	1012	1322342.20004	093300.07901	-0.00	1.01	1.40
Centroid	1326543.44799	689875.71779		1326543.44799	689875.71779	0.00	-0.00	6.06

Centroid to Translated Point

	Latitude	Departure	D/L	ATAN	DEGREES	Azimuth	Distance	GON
1027	5286.59683	-3864.61760	-0.73102	-0.63124	-36.16762	323.83238	6548.54	359.81376
1023	2864.21504	-2277.50898	-0.79516	-0.67178	-38.49031	321.50969	3659.34	357.23299
1019	156.10506	-197.41636	-1.26464	-0.90173	-51.66516	308.33484	251.68	342.59427
1017	-116.05803	46.12504	-0.39743	-0.37829	-21.67440	158.32560	124.89	175.91733
1020	22.73786	-346.45726	-15.23702	-1.50526	-86.24508	273.75492	347.20	304.17213
1018	-104.33491	-232.74794	2.23078	1.14938	65.85454	265.85454	255.06	273.17171
1011	-3908.07743	3439.36149	-0.88006	-0.72189	-41.34957	138.65013	5205.98	154.05570
1012	-4201.18445	3433.26158	-0.81721	-0.68515	-39.25313	140.74387	5425.61	156.38207

3 Parameter - Rotation and Translation of calculated positions

	Found Monumentation			Translated Coordinates				
	Northing	Easting		Northing	Easting	Northing Diff	Easting Diff	Residual
1080	1331830.09268	686010.51712	1027	1331829.91429	686010.54809	0.18	-0.03	0.18
395	1329407.53410	687597.74156	1023	1329407.66824	687597.86387	-0.13	-0.12	0.18
393	1326700.33016	689677.77756	1019	1326699.73617	689678.18808	0.59	-0.41	0.72
1222	1326427.66494	689921.80973	1017	1326427.59391	689921.75276	0.07	0.06	0.09
407	1326566.90762	689528.70143	1020	1326566.35622	689529.15859	0.55	-0.46	0.72
1220	1326439.19739	689643.10713	1018	1326439.29318	689642.87878	-0.10	0.23	0.25
1216	1322634.53973	693316.04206	1011	1322635.86471	693315.31349	-1.32	0.73	1.51
1215	1322341.31727	693310.04572	1012	1322342.75717	693309.23865	-1.44	0.81	1.65
Centroid	1326543.44799	689875.71779		1326543.64799	689875.61779	-0.20	0.10	5.30
Rotation	-0.00490							
Shift	0.20	-0.10						

= | SIN ▼ RADIANS ▼ (Azimuth 1027 ▼) + $Northing $Rotation ▼ × ⊗ ✓

Distance 1027 ▼ + Synchronized Centroids::$F$12 ▼ + $C$14 ▼

Translated Easting = Sin(toRadians(Azimuth + Rotation)) x
Distance + Centroid Easting + Easting Shift

Repeat for each point in the data set.

 When using Azimuth, the Cos and Sin values automatically add the appropriate algebraic sign to the solution. The Rotational value is in d.ddd format.

Create two new columns for the Northing and Easting Difference.

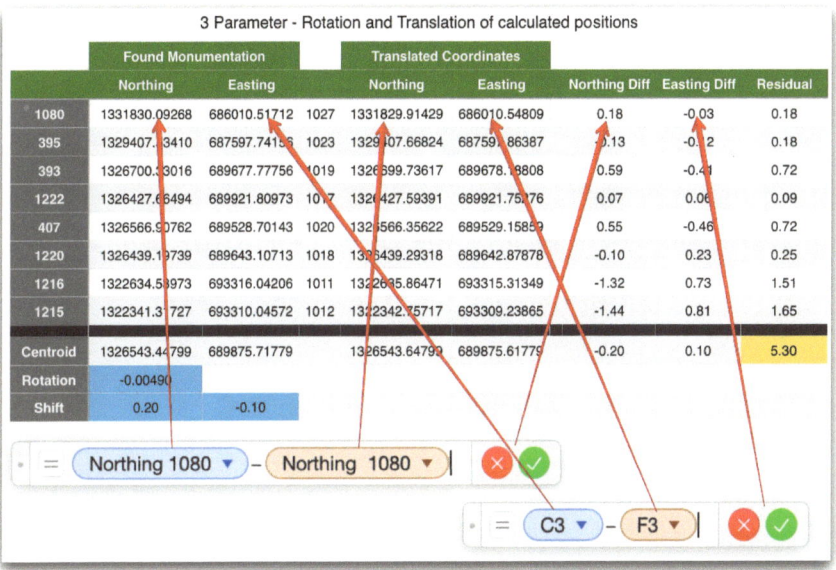

3 Parameter - Rotation and Translation of calculated positions

	Found Monumentation			Translated Coordinates				
	Northing	Easting		Northing	Easting	Northing Diff	Easting Diff	Residual
1080	1331830.09268	686010.51712	1027	1331829.91429	686010.54809	0.18	-0.03	0.18
395	1329407.3410	687597.74156	1023	1329407.66824	687591.86387	0.13	-0.12	0.18
393	1326700.33016	689677.77756	1019	1326699.73617	689678.18808	0.59	-0.41	0.72
1222	1326427.66494	689921.80973	1017	1326427.59391	689921.75176	0.07	0.06	0.09
407	1326566.90762	689528.70143	1020	1326566.35622	689529.15859	0.55	-0.46	0.72
1220	1326439.19739	689643.10713	1018	1326439.29318	689642.87878	-0.10	0.23	0.25
1216	1322634.53973	693316.04206	1011	1322635.86471	693315.31349	-1.32	0.73	1.51
1215	1322341.31727	693310.04572	1012	1322342.75717	693309.23865	-1.44	0.81	1.65
Centroid	1326543.44799	689875.71779		1326543.64799	689875.61779	-0.20	0.10	5.30
Rotation	-0.00490							
Shift	0.20	-0.10						

• = (Northing 1080 ▼) − (Northing 1080 ▼) ✕ ✓

• = (C3 ▼) − (F3 ▼) ✕ ✓

Northing Diff = Found Northing - Translated Northing
Easting Diff = Found Easting - Translated Easting

Repeat for each point in the data set.

Create a new column for the Residual.

Crume's Transformation

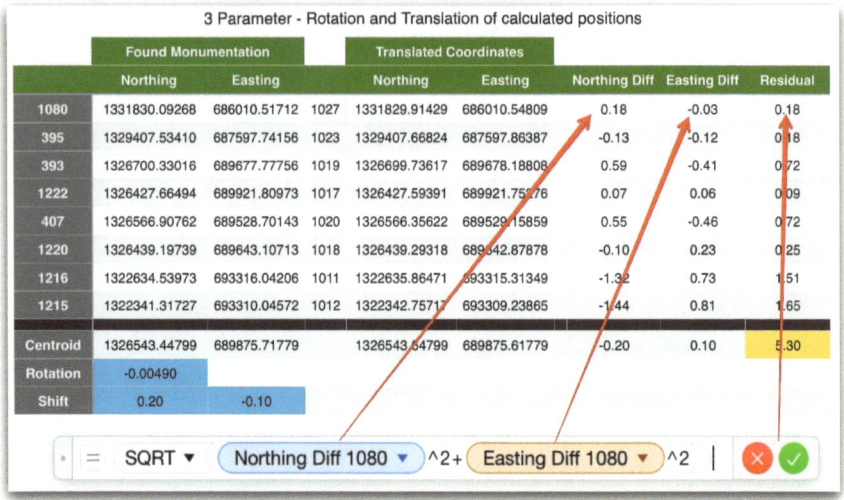

3 Parameter - Rotation and Translation of calculated positions

	Found Monumentation			Translated Coordinates				
	Northing	Easting		Northing	Easting	Northing Diff	Easting Diff	Residual
1080	1331830.09268	686010.51712	1027	1331829.91429	686010.54809	0.18	-0.03	0.18
395	1329407.53410	687597.74156	1023	1329407.66824	687597.86387	-0.13	-0.12	0.18
393	1326700.33016	689677.77756	1019	1326699.73617	689678.18808	0.59	-0.41	0.72
1222	1326427.66494	689921.80973	1017	1326427.59391	689921.75176	0.07	0.06	0.09
407	1326566.90762	689528.70143	1020	1326566.35622	689529.15859	0.55	-0.46	0.72
1220	1326439.19739	689643.10713	1018	1326439.29318	689642.87878	-0.10	0.23	0.25
1216	1322634.53973	693316.04206	1011	1322635.86471	693315.31349	-1.32	0.73	1.51
1215	1322341.31727	693310.04572	1012	1322342.75717	693309.23865	-1.44	0.81	1.65
Centroid	1326543.44799	689875.71779		1326543.64799	689875.61779	-0.20	0.10	5.30
Rotation	-0.00490							
Shift	0.20	-0.10						

$\bullet$ = SQRT ▼ (Northing Diff 1080 ▼)^2+ (Easting Diff 1080 ▼)^2 | ❌ ✅

Residual = SQRT(Northing Diff^2 + Easting Diff^2)

Repeat for each point in the data set.

Create a row for the Centroid and summation of the Northing Diff, Easting Diff and Residual.

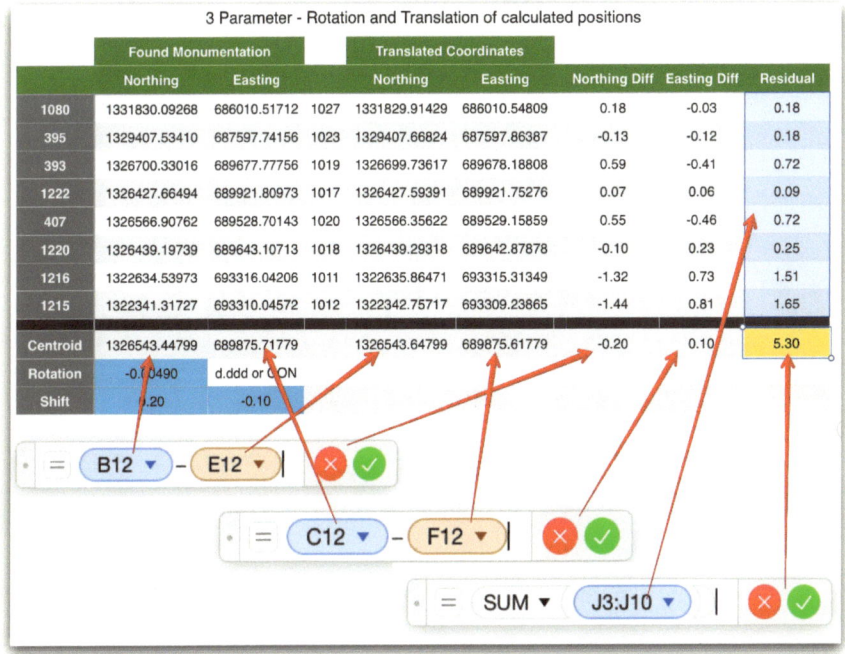

3 Parameter - Rotation and Translation of calculated positions

	Found Monumentation			Translated Coordinates		Northing Diff	Easting Diff	Residual
	Northing	Easting		Northing	Easting			
1080	1331830.09268	686010.51712	1027	1331829.91429	686010.54809	0.18	-0.03	0.18
395	1329407.53410	687597.74156	1023	1329407.66824	687597.86387	-0.13	-0.12	0.18
393	1326700.33016	689677.77756	1019	1326699.73617	689678.18808	0.59	-0.41	0.72
1222	1326427.66494	689921.80973	1017	1326427.59391	689921.75276	0.07	0.06	0.09
407	1326566.90762	689528.70143	1020	1326566.35622	689529.15859	0.55	-0.46	0.72
1220	1326439.19739	689643.10713	1018	1326439.29318	689642.87878	-0.10	0.23	0.25
1216	1322634.53973	693316.04206	1011	1322635.86471	693315.31349	-1.32	0.73	1.51
1215	1322341.31727	693310.04572	1012	1322342.75717	693309.23865	-1.44	0.81	1.65
Centroid	1326543.44799	689875.71779		1326543.64799	689875.61779	-0.20	0.10	5.30
Rotation	-0.0490	d.ddd or CON						
Shift	0.20	-0.10						

= B12 ▼ - E12 ▼ | ✕ ✓

= C12 ▼ - F12 ▼ | ✕ ✓

= SUM ▼ J3:J10 ▼ | ✕ ✓

Northing Diff = (Centroid) Found Northing - Translated Northing

Easting Diff = (Centroid) Found Easting - Translated Easting

Residual = Sum of all points

Create a new table named "Standard Deviation".

Add columns for the Northing Diff, Easting Diff and Residual.

Crume's Transformation

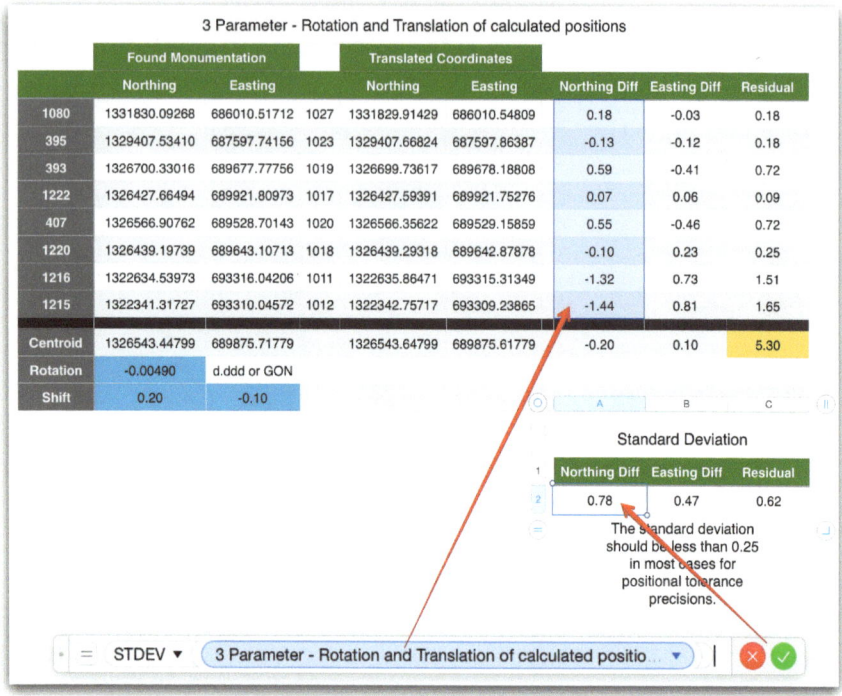

3 Parameter - Rotation and Translation of calculated positions

	Found Monumentation			Translated Coordinates		Northing Diff	Easting Diff	Residual
	Northing	Easting		Northing	Easting			
1080	1331830.09268	686010.51712	1027	1331829.91429	686010.54809	0.18	-0.03	0.18
395	1329407.53410	687597.74156	1023	1329407.66824	687597.86387	-0.13	-0.12	0.18
393	1326700.33016	689677.77756	1019	1326699.73617	689678.18808	0.59	-0.41	0.72
1222	1326427.66494	689921.80973	1017	1326427.59391	689921.75276	0.07	0.06	0.09
407	1326566.90762	689528.70143	1020	1326566.35622	689529.15859	0.55	-0.46	0.72
1220	1326439.19739	689643.10713	1018	1326439.29318	689642.87878	-0.10	0.23	0.25
1216	1322634.53973	693316.04206	1011	1322635.86471	693315.31349	-1.32	0.73	1.51
1215	1322341.31727	693310.04572	1012	1322342.75717	693309.23865	-1.44	0.81	1.65
Centroid	1326543.44799	689875.71779		1326543.64799	689875.61779	-0.20	0.10	5.30
Rotation	-0.00490	d.ddd or GON						
Shift	0.20	-0.10						

Standard Deviation

Northing Diff	Easting Diff	Residual
0.78	0.47	0.62

The Standard deviation should be less than 0.25 in most cases for positional tolerance precisions.

= STDEV ▼ (3 Parameter - Rotation and Translation of calculated positio ▼) | ✕ ✓

Repeat the formula for each of the columns.

THE FUN BEGINS

With the tables being built, you can now start evaluating the rotational component with a Northing and Easting Shift.

The Goal is to minimize the "Sum of the Residuals" along with minimizing the "Standard Deviation".

Start by inputing a small rotational value both clockwise and counter clockwise until you find the minimum value for the "Sum of the Residuals".

Next apply a shift value to the Northing and Easting and monitor the "Sum of the Residuals".

If the found monuments were in exactly the same position as the record geometry, the "Sum of the Residuals" would be Zero. That is never going to happen. The goal is to first check to see if the found monuments are in a close geometric relationship with the record by reviewing all of the residuals. The smaller the residuals, the closer the geometric relationship.

Next review the Residual amounts to see if there might be one or two found monuments that just aren't playing well with the rest of the group. You might want to remove those monuments from the data set and see what happens.

In our example, points 1216 & 1215 do not fit well with the others based upon the higher residuals.

Crume's Transformation

3 Parameter - Rotation and Translation of calculated positions

	Found Monumentation			Translated Coordinates				
	Northing	Easting		Northing	Easting	Northing Diff	Easting Diff	Residual
1080	1331830.09268	686010.51712	1027	1331829.91429	686010.54809	0.18	-0.03	0.18
395	1329407.53410	687597.74156	1023	1329407.66824	687597.86387	-0.13	-0.12	0.18
393	1326700.33016	689677.77756	1019	1326699.73617	689678.18808	0.59	-0.41	0.72
1222	1326427.66494	689921.80973	1017	1326427.59391	689921.75276	0.07	0.06	0.09
407	1326566.90762	689528.70143	1020	1326566.35622	689529.15859	0.55	-0.46	0.72
1220	1326439.19739	689643.10713	1018	1326439.29318	689642.87878	-0.10	0.23	0.25
Centroid	1327895.28782	688729.94242		1327895.09367	688730.06503	0.19	-0.12	2.14
Rotation	-0.00490	d.ddd or GON						
Shift	0.20	-0.10						

Standard Deviation

Northing Diff	Easting Diff	Residual
0.31	0.27	0.29

The standard deviation should be less than 0.25 in most cases for positional tolerance precisions.

By removing them, the "Sum of the Residuals" and "Standard Deviation" have improved. You can change the Rotational and Shift values to see what effect that has on the solution.

Once you have determined the "Best Fit" solution with the found monuments, adjust the calculated record centerline coordinates by the same Centroid Translation value, the Rotation value and Shift value in that order.

Per the example:

Centroid Translation value = (N 3.57) (E -16.03)

Rotation value = -0.00490 (Origin is the Centroid)

Northing and Easting Shift value = (N 0.20) (E -0.10)

Once the centerline has been updated, you can perform the final right of way calculations.

GONS

The use of GON is very similar to using Azimuths. Replace the Azimuth formulas above with these formulas.

You first convert GON to Radians then everything else is the same as using Azimuths.

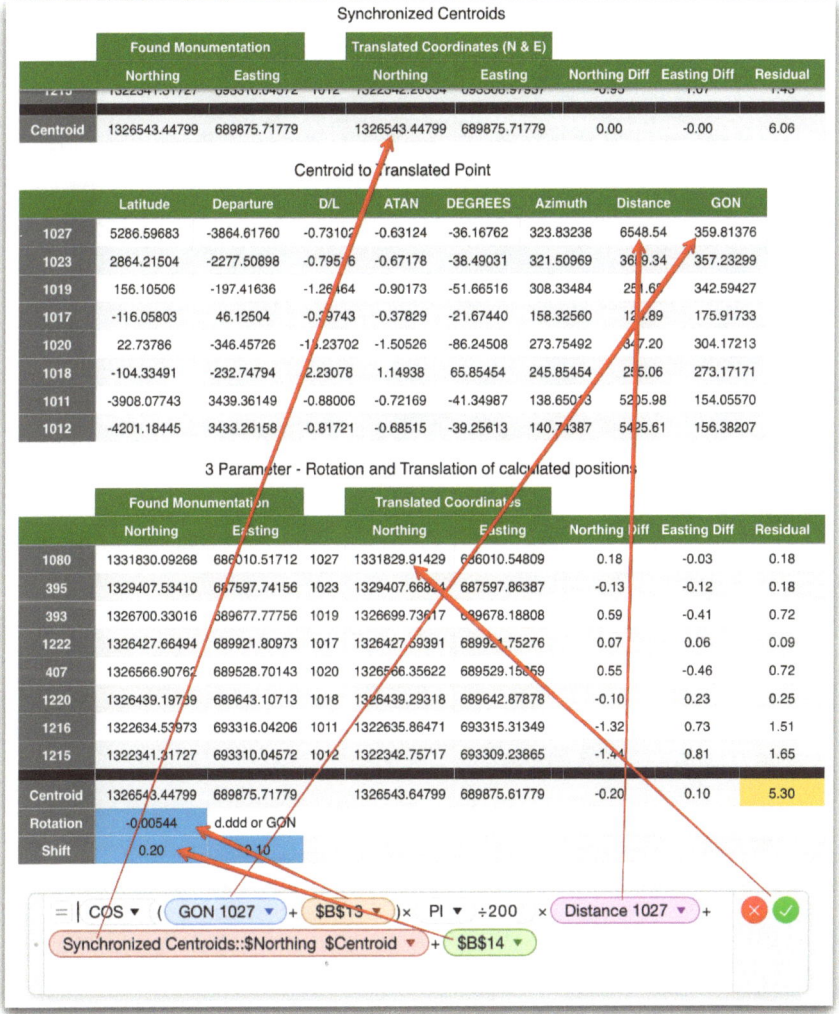

Synchronized Centroids

| | Found Monumentation | | | Translated Coordinates (N & E) | | | | |
	Northing	Easting		Northing	Easting	Northing Diff	Easting Diff	Residual
1215	1322341.31727	693310.04572	1012	1322342.20304	693306.97957	-0.95	1.07	1.45
Centroid	1326543.44799	689875.71779		1326543.44799	689875.71779	0.00	-0.00	6.06

Centroid to Translated Point

	Latitude	Departure	D/L	ATAN	DEGREES	Azimuth	Distance	GON
1027	5286.59683	-3864.61760	-0.73102	-0.63124	-36.16762	323.83238	6548.54	359.81376
1023	2864.21504	-2277.50898	-0.79516	-0.67178	-38.49031	321.50969	3659.34	357.23299
1019	156.10506	-197.41636	-1.26464	-0.90173	-51.66516	308.33484	251.69	342.59427
1017	-116.05803	46.12504	-0.39743	-0.37829	-21.67440	158.32560	124.89	175.91733
1020	22.73786	-346.45726	-15.23702	-1.50526	-86.24508	273.75492	347.20	304.17213
1018	-104.33491	-232.74794	2.23078	1.14938	65.85454	245.85454	255.06	273.17110
1011	-3908.07743	3439.36149	-0.88006	-0.72169	-41.34987	138.65073	5205.98	154.05570
1012	-4201.18445	3433.26158	-0.81721	-0.68515	-39.25613	140.74387	5425.61	156.38207

3 Parameter - Rotation and Translation of calculated positions

| | Found Monumentation | | | Translated Coordinates | | | | |
	Northing	Easting		Northing	Easting	Northing Diff	Easting Diff	Residual
1080	1331830.09268	686010.51712	1027	1331829.91429	686010.54809	0.18	-0.03	0.18
395	1329407.53410	687597.74156	1023	1329407.66814	687597.86387	-0.13	-0.12	0.18
393	1326700.33016	689677.77756	1019	1326699.73617	689678.18808	0.59	-0.41	0.72
1222	1326427.66494	689921.80973	1017	1326427.59391	689921.75276	0.07	0.06	0.09
407	1326566.90762	689528.70143	1020	1326566.35622	689529.15859	0.55	-0.46	0.72
1220	1326439.19739	689643.10713	1018	1326439.29318	689642.87878	-0.10	0.23	0.25
1216	1322634.53973	693316.04206	1011	1322635.86471	693315.31349	-1.32	0.73	1.51
1215	1322341.31727	693310.04572	1012	1322342.75717	693309.23865	-1.44	0.81	1.65
Centroid	1326543.44799	689875.71779		1326543.64799	689875.61779	-0.20	0.10	5.30
Rotation	-0.00544	d.ddd or GON						
Shift	0.20	0.10						

= | COS ▼ (GON 1027 ▼ + $B$13 ▼)× PI ▼ ÷200 × Distance 1027 ▼ +

Synchronized Centroids::$Northing $Centroid ▼ + $B$14 ▼

$$\text{Translated Northing} = \text{Cos}((\text{GON} + \text{Rotation}) * \text{PI} / 200) \text{ x}$$
$$\text{Distance} + \text{Centroid Northing} + \text{Northing Shift}$$

Repeat for each point in the data set.

Synchronized Centroids

	Found Monumentation			Translated Coordinates (N & E)		Northing Diff	Easting Diff	Residual
	Northing	Easting		Northing	Easting			
1215	1322341.31727	693310.04572	1012	1322342.26354	693308.97937	-0.95	1.07	1.43
Centroid	1326543.44799	689875.71779		1326543.44799	689875.71779	0.00	-0.00	6.06

Centroid to Translated Point

	Latitude	Departure	D/L	ATAN	DEGREES	Azimuth	Distance	GON
1027	5286.59683	-3864.61760	-0.73102	-0.63124	36.16762	323.83238	6548.54	359.81376
1023	2864.21504	-2277.50898	-0.79516	-0.67178	-38.49031	321.50969	3653.34	357.23299
1019	156.10506	-197.41636	-1.26464	-0.9017?	-51.66516	308.33484	25?.68	342.59427
1017	-116.05803	46.12504	-0.39743	-0.37?29	-21.67440	158.32560	12?.?9	175.91733
1020	22.73786	-346.45726	-15.23702	-1.?0526	-86.24508	273.75492	3?7.20	304.17213
1018	-104.33491	-232.74794	2.23078	?.14938	65.85454	245.85454	2?5.06	273.17171
1011	-3908.07743	3439.36149	-0.88006	-0.72169	-41.34987	138.6501?	52?5.98	154.05570
1012	-4201.18445	3433.26158	-0.8172?	-0.68515	-39.25613	140.74?87	54?5.61	156.38207

3 Parameter - Rotation and Translation of calculated positions

	Found Monumentation			Translated Coordinates		Northing Diff	Easting Diff	Residual
	Northing	Easting		Northing	Easting			
1080	1331830.09268	686010.517?2	1027	1331829.91429	686010.54809	0.18	-0.03	0.18
395	1329407.53410	687597.7?156	1023	1329407.66824	687597.86387	-0.13	-0.12	0.18
393	1326700.33016	689677?77756	1019	1326699.736?7	689678.18808	0.59	-0.41	0.72
1222	1326427.66494	689921.80973	1017	1326427.59391	689921.75276	?.07	0.06	0.09
407	1326566.90762	689528.70143	1020	1326566.35622	689529.15859	0.5?	-0.46	0.72
1220	1326439.19739	689643.10713	1018	1326439.29318	689642.87878	-0.10	0.23	0.25
1216	1322634.53973	693316.04206	1011	1322635.86471	693315.31349	-1.32	0.73	1.51
1215	1322341.31727	693310.04572	1012	1322342.75717	693309.23865	-1.44	0.81	1.65
Centroid	1326543.44799	689875.71779		1326543.64799	689875.61779	-0.20	0.10	5.30
Rotation	-0.00544	d.ddd or GON						
Shift	0.20	-0.10						

= | SIN ▼ ((GON 1027 ▼ + $B$13 ▼) x PI ▼ ÷200 x (Distance 1027 ▼ +
Synchronized Centroids::$F$12 ▼ + $C$14 ▼

$$\text{Translated Easting} = \text{Sin}((\text{GON} + \text{Rotation}) * \text{PI} / 200) \text{ x}$$
$$\text{Distance} + \text{Centroid Easting} + \text{Easting Shift}$$

Repeat for each point in the data set.

CRUME'S TRANSFORMATION

Create record Positions

Calculate the Centroids

Sychronize the Centroids

Rotate Record Positions

Shift N-S, E-W

Repeat
(Rotation and Shift as needed)
to Minimize the Sum of
Residuals

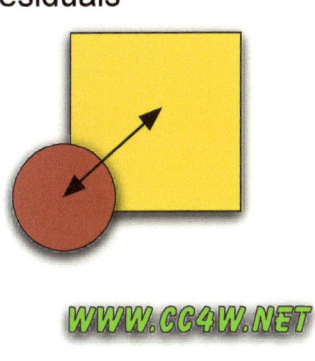

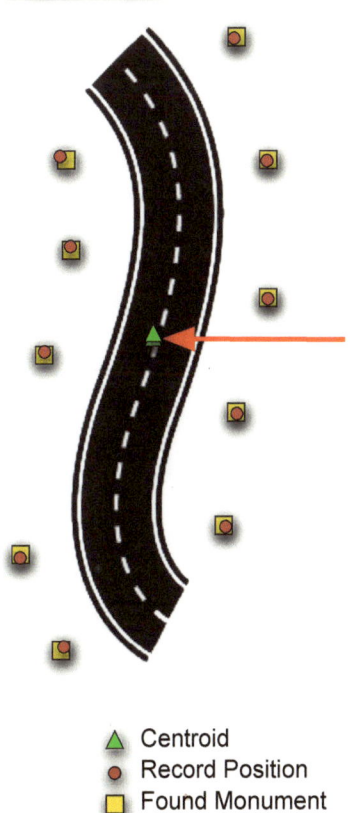

WWW.CC4W.NET

▲ Centroid
● Record Position
☐ Found Monument

There is a simple video that demonstrates the Fixed Points Transformation locate at:

https://youtu.be/AGFQvVhe4Mg

RANDOM POINTS METHOD

This method was designed for holding record geometry through a set of random points such as fences, edge of pavement, lane striping, centerline striping.

Use this method when there are no existing monumentation, section line ties, existing structures (bridges, culverts, etc.) or other fixed positions available.

First start by plotting the random points in CADD per the measured surveyed points. For our example the right of way fence line was located at random intervals on each side of the highway.

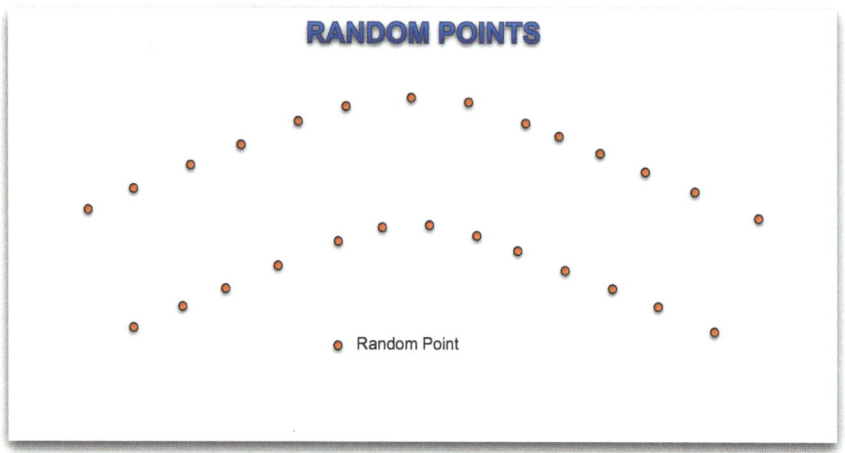

Next determine a starting point for the record centerline alignment for the survey limits. This point will be a split between the left and right random point at the beginning of the centerline alignment.

From this point, calculate the record centerline alignment using a coordinate geometry program and CADD. We will be adjusting the centerline alignment to a "Best Fit" with the random points as we continue with this transformation.

Crume's Transformation

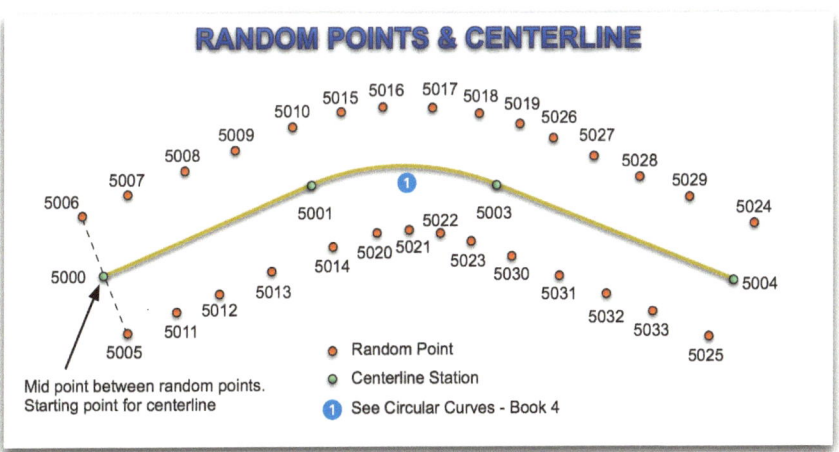

RANDOM POINTS & CENTERLINE

Mid point between random points.
Starting point for centerline

- Random Point
- Centerline Station
- ① See Circular Curves - Book 4

Centerline Alignment Report

Project: Book 14 Highway Centerline

Line Name: CL-Record Geometry
Course Data

Point No.	Direction	Distance	Northing	Easting
5000	*(Sta) 10+00.00*		1320960.98830	693294.84033
	N 68-22-10 E	1500.000		
5001	*(Sta) 25+00.00*		1321513.91882	694689.21038
********** 5001 (PC)				
Circular Curve Data		RT		
5002 (RP)			1319654.75875	695426.45108
	L.T.B. (in):	N 68-22-10 E		
	D.O.C. Arc:	02-51-53		
	Radius:	2000.00		
	Delta Angle:	45-00-00 (RT)		
	Tangent length:	828.43		
	Arc length:	1570.796		
	L.T.B. (out):	S 66-37-50 E		
********** 5003 (PT)				
5003	*(Sta) 40+70.80*		1321490.69133	696219.76788
	S 66-37-50 E	1800.000		
5004	*(Sta) 58+70.80*		1320776.70621	697872.10720

(SimPro - custom software by Creative Computing 4 Windows) www.cc4w.net
(Licensed to Creative Computing 4 Windows)

33

Next create a station and offset report for the random points using the record centerline alignment.

Station-Offset Report				

Project: Book 14 Highway Centerline

Report Name: Sta-Off Random
Centerline: CL-Record Geometry

Point No.	Station	Offset	Northing	Easting
[--Tangent-- 5000 to 5001]				
5005	P.O.T. 10+01.45	99.840 RT	1320868.71412	693332.99382
5007	P.O.T. 12+70.28	99.360 LT	1321152.98412	693509.46537
5008	P.O.T. 16+66.20	99.338 LT	1321298.90724	693877.51183
5009	P.O.T. 20+90.66	99.215 LT	1321455.25540	694272.12270
5011	P.O.T. 12+84.00	99.923 RT	1320972.78915	693595.67127
5012	P.O.T. 17+26.56	100.051 RT	1321135.80625	694007.11406
5013	P.O.T. 21+59.48	100.084 RT	1321295.35922	694409.56123
[--Curve-- 5001-PC 5002RP 5003-PT]				
5010	P.O.C. 25+01.58	99.096 LT	1321606.64677	694654.22302
5014	P.O.C. 25+01.53	100.276 RT	1321421.23901	694727.52465
5015	P.O.C. 27+83.85	99.150 LT	1321695.91891	694936.45578
5016	P.O.C. 30+84.45	99.414 LT	1321746.53916	695247.58611
5017	P.O.C. 33+62.31	99.669 LT	1321751.40932	695539.00343
5018	P.O.C. 37+14.81	99.960 LT	1321699.47681	695904.94754
5020	P.O.C. 28+32.88	100.215 RT	1321512.37677	695028.40744
5021	P.O.C. 31+84.74	99.930 RT	1321553.65483	695359.66590
5022	P.O.C. 36+18.99	99.779 RT	1321523.60403	695770.33412
[--Tangent-- 5003 to 5004]				
5019	P.O.T. 40+72.43	99.954 LT	1321581.79635	696260.91858
5023	P.O.T. 40+72.53	99.413 RT	1321398.74402	696181.93043
5026	P.O.T. 44+02.06	99.858 LT	1321450.95876	696563.46759
5027	P.O.T. 46+56.90	99.784 LT	1321349.80686	696797.37138
5028	P.O.T. 50+10.29	99.590 LT	1321209.45366	697121.69309
5029	P.O.T. 54+51.86	99.553 LT	1321034.26528	697527.03018
5030	P.O.T. 43+77.78	99.502 RT	1321277.58256	696462.10437
5031	P.O.T. 46+47.84	99.672 RT	1321170.30584	696709.93997
5032	P.O.T. 49+82.99	99.678 RT	1321037.35991	697017.59554
5033	P.O.T. 53+68.89	99.698 RT	1320884.27211	697371.82724

Points with no solution:
5006
5024
5025

Create a table in Numbers (Mac) or Excel (Windows) or other spreadsheet program.

Place the offset distances for each random point in a column named "1st".

	1st	2nd	3rd	4th
5005	99.84			
5007	99.36			
5008	99.34			
5009	99.22			
5010	99.10			
5011	99.92			
5012	100.05			
5013	100.08			
5014	100.28			
5015	99.15			
5016	99.41			
5017	99.67			
5018	99.96			
5019	99.95			
5020	100.22			
5021	99.93			
5022	99.78			
5023	99.41			
5026	99.86			
5027	99.78			
5028	99.59			
5029	99.55			
5030	99.50			
5031	99.67			
5032	99.68			
5033	99.70			

As we progress through the transformation there will be several iterations that will be processed and reviewed.

Create a table named "Mean & Standard Deviation".

Calculate the "Mean" of the offsets in a cell. Calculate the "Standard Deviation" of the offsets in a separate cell.

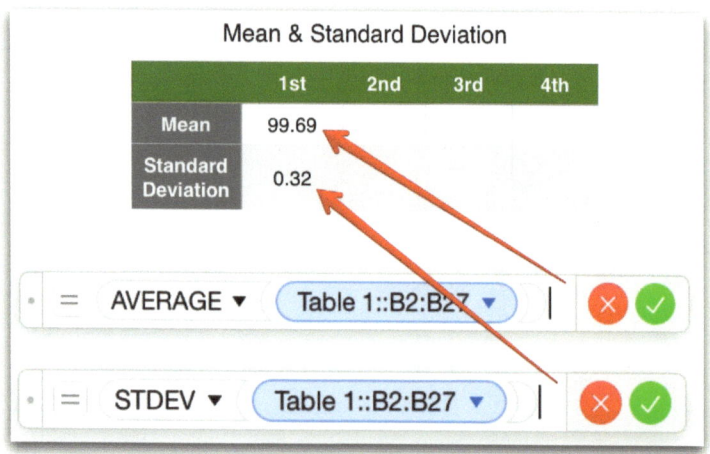

For this first iteration the "Mean" and "Standard Deviation" will be high. This is to be expected since you are establishing a baseline to start the adjustments from.

If the record centerline was perfectly aligned with the random points and they were set at the correct distance the "Mean" would be 99.50' (assuming that the fence was set 0.50' inside of the 100' right of way width left and right) and the "Standard Deviation" would be 0.00 meaning that there is no deviation from the Mean.

In the real world the "Mean" will never be exactly 99.50' (for this example) and the "Standard Deviation" will never be exactly 0.00, however the goal is to find the minimum value for each.

Keep in mind that whatever you are using, a fence, edge of pavement, lane striping, centerline striping, etc., they will be at random distances from the controlling centerline due to how they were constructed. We are merely trying to find a mathematical "Best Fit" within this randomness.

When using GPS to collect random points, they will be based upon Grid Bearings. It is assumed that a calibration/localization was not performed on multiple control points. Calibration/localization is a topic all on it's own and will not be covered in this book.

If you are working in GONS, the record direction datum will probably be different than the GPS projection you are working in. You will need to verify this.

The record centerline alignment will need to be rotated by some amount and shifted in a North-South and/or East-West direction. With each iteration these adjustments will be made in an effort to minimize the "Mean" (99.50 for our example) and "Standard Deviation" (0.00). For centerline striping, the desirable "Mean" would be 0.00. For edge of pavement, fencing, lane striping, etc. the desirable "Mean" would be the offset distance from the centerline.

The record centerline alignment may or may not be on Grid Bearings and more than likely it will need to be rotated to get the alignment to fit closer to the random points.

To determine the rotation value first review the offsets for each of the tangent lines along the record centerline alignment.

Select a set of random points that are opposite each other that are towards the beginning of the tangent line and at the end of the tangent line.

In CADD connect the opposite random points with a line. From the mid-point of each of these lines draw a line connecting the mid-points.

In CADD measure the angle between the record centerline alignment and the line that connects the mid-points.

See below for the measured angle of the two tangent lines of our example.

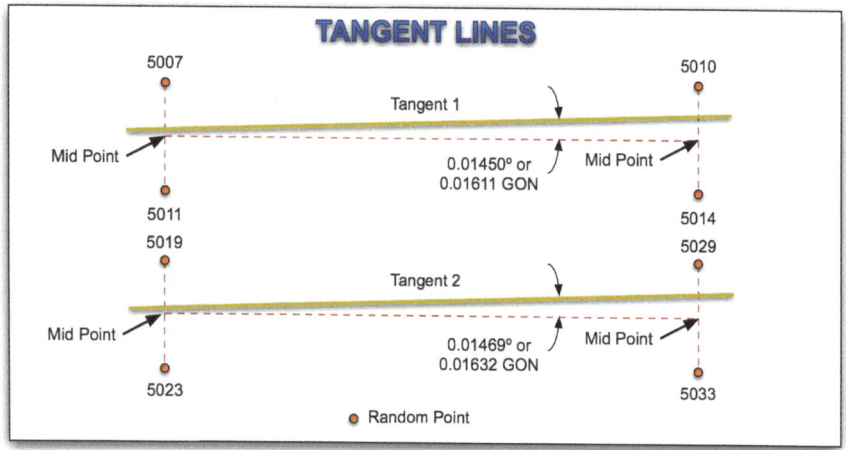

Review the measured angles to see how close they are to each other. Pick one or average the two to use as the rotation value for the record centerline.

Rotate the record centerline alignment by the measured angle holding the beginning point (5000) of the centerline as the pivot point then create a new station and offset report for the random points.

Place the offset distances for each random point in a column named "2nd".

Crume's Transformation

	1st	2nd	3rd	4th
5005	99.84	99.84		
5007	99.36	99.43		
5008	99.34	99.51		
5009	99.22	99.49		
5010	99.10	99.48		
5011	99.92	99.85		
5012	100.05	99.87		
5013	100.08	99.79		
5014	100.28	99.89		
5015	99.15	99.60		
5016	99.41	99.93		
5017	99.67	100.23		
5018	99.96	100.57		
5019	99.95	100.59		
5020	100.22	99.75		
5021	99.93	99.40		
5022	99.78	99.18		
5023	99.41	98.78		
5026	99.86	100.59		
5027	99.78	100.57		
5028	99.59	100.47		
5029	99.55	100.54		
5030	99.50	98.79		
5031	99.67	98.89		
5032	99.68	98.10		
5033	99.70	98.73		

Add the "Mean" and "Standard Deviation" for the "2nd" column.

Mean & Standard Deviation				
	1st	2nd	3rd	4th
Mean	99.69	99.69		
Standard Deviation	0.32	0.67		

As you can see the "Mean" is holding steady but the "Standard Deviation" has increased.

Next let's look at the graphic linework to determine how much a North-South and East-West shift might help improve the "Best Fit".

The best place to look is in the middle of the centerline alignment near a PC or PT.

In CADD zoom into the area of point 5003. The random points 5019 and 5023 appear to be close to the PT (5003) of the curve.

From the mid-point between these two random points draw a horizontal line. From the PT of the curve draw a vertical line.

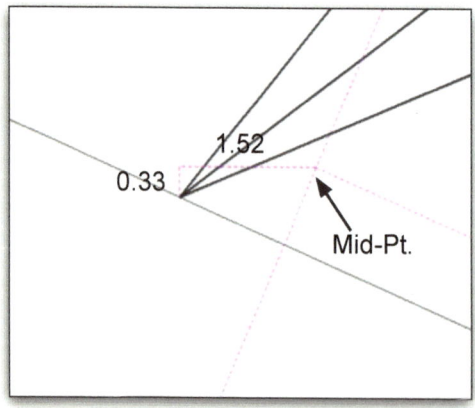

Measure the North-South distance (0.33) and the East-West distance (1.52).

Move the centerline alignment by 0.33 North and 1.52 East then run a new station and offset report.

Add the offsets to the "3rd" column and create a "Mean" and "Standard Deviation" for this column.

Crume's Transformation

	1st	2nd	3rd	4th
5005	99.84	99.84	99.68	
5007	99.36	99.43	99.76	
5008	99.34	99.51	99.75	
5009	99.22	99.49	99.75	
5010	99.10	99.48	99.74	
5011	99.92	99.85	99.60	
5012	100.05	99.87	99.61	
5013	100.08	99.79	99.53	
5014	100.28	99.89	99.64	
5015	99.15	99.60	99.64	
5016	99.41	99.93	99.73	
5017	99.67	100.23	99.82	
5018	99.96	100.57	99.90	
5019	99.95	100.59	99.68	
5020	100.22	99.75	99.75	
5021	99.93	99.40	99.67	
5022	99.78	99.18	99.78	
5023	99.41	98.78	99.68	
5026	99.86	100.59	99.62	
5027	99.78	100.57	99.66	
5028	99.59	100.47	99.56	
5029	99.55	100.54	99.64	
5030	99.50	98.79	99.69	
5031	99.67	98.89	99.80	
5032	99.68	98.10	99.72	
5033	99.70	98.73	99.64	

Mean & Standard Deviation

	1st	2nd	3rd	4th
Mean	99.69	99.69	99.69	
Standard Deviation	0.32	0.67	0.08	

The "Mean" is holding steady and the "Standard Deviation" has decreased substantially.

The "Mean" for a perfect fit would be 99.50 if the fence was constructed exactly at 99.50 feet from the centerline. The Mean is 99.69 which is within 0.19'. This is pretty good considering the fence was not constructed perfectly.

The "3rd" iteration is looking pretty good as a "Best Fit". As a check let's move the centerline a little bit more North and East then review the Mean and Standard Deviation.

For the "4th" iteration let's move the centerline North 0.50 and East 0.50 for a check.

Add the offsets to the "4th" column and create a "Mean" and "Standard Deviation" for this column.

Crume's Transformation

	1st	2nd	3rd	4th
5005	99.84	99.84	99.68	
5007	99.36	99.43	99.76	99.40
5008	99.34	99.51	99.75	99.48
5009	99.22	99.49	99.75	99.47
5010	99.10	99.48	99.74	99.45
5011	99.92	99.85	99.60	99.88
5012	100.05	99.87	99.61	99.89
5013	100.08	99.79	99.53	99.81
5014	100.28	99.89	99.64	99.92
5015	99.15	99.60	99.64	99.27
5016	99.41	99.93	99.73	99.28
5017	99.67	100.23	99.82	99.30
5018	99.96	100.57	99.90	99.30
5019	99.95	100.59	99.68	99.02
5020	100.22	99.75	99.75	100.14
5021	99.93	99.40	99.67	100.15
5022	99.78	99.18	99.78	100.36
5023	99.41	98.78	99.68	100.34
5026	99.86	100.59	99.62	99.02
5027	99.78	100.57	99.66	99.01
5028	99.59	100.47	99.56	98.90
5029	99.55	100.54	99.64	98.98
5030	99.50	98.79	99.69	100.35
5031	99.67	98.89	99.80	100.45
5032	99.68	98.10	99.72	100.37
5033	99.70	98.73	99.64	100.29

Mean & Standard Deviation

	1st	2nd	3rd	4th
Mean	99.69	99.69	99.69	99.67
Standard Deviation	0.32	0.67	0.08	0.53

The "Mean" decreased slightly however the "Standard Deviation" increased substantially.

The "3rd" iteration is by far the best solution so far. Additional iterations can be run by moving the centerline by smaller increments until the "Best Fit" is achieved. For our example we are going to use the "3rd" iteration as the "Best Fit" solution.

Below is the final "Best Fit" alignment. You can now perform the final right of way calculations.

Centerline Alignment Report

Project: Book 14 Highway Centerline

Line Name: CL-EX RW - Best Fit

Course Data

Point No.	Direction	Distance	Northing	Easting
5000	*(Sta) 10+00.00*		1320961.31709	693296.36125
	N 68-23-03 E	1500.000		
5001	*(Sta) 25+00.00*		1321513.89009	694690.87302
********** 5001 (PC)				
Circular Curve Data		RT		
5002 (RP)			1319654.54106	695427.63702
	L.T.B. (in):	N 68-23-03 E		
	D.O.C. Arc:	02-51-53		
	Radius:	2000.00		
	Delta Angle:	45-00-00 (RT)		
	Tangent length:	828.43		
	Arc length:	1570.796		
	L.T.B. (out):	S 66-36-57 E		
********** 5003 (PT)				
5003	*(Sta) 40+70.80*		1321490.27018	696221.42450
	S 66-36-57 E	1800.000		
5004	*(Sta) 58+70.80*		1320775.86145	697873.58072

(**SimPro** - custom software by Creative Computing 4 Windows) www.cc4w.net
(Licensed to Creative Computing 4 Windows)

ABOUT THE AUTHOR

Jim Crume P.L.S., M.S., CFedS

My land surveying career began several decades ago while attending Albuquerque Technical Vocational Institute in New Mexico and has traversed many states such as Alaska, Arizona, Utah and Wyoming. I am a Professional Land Surveyor in Arizona, Utah and Wyoming. I am an appointed United States Mineral Surveyor and a Bureau of Land Management (BLM) Certified Federal Surveyor. I have many years of computer programming experience related to surveying.

This ebook is dedicated to the many individuals that have helped shape my career. Especially my wife Cindy. She has been my biggest supporter. She has been my instrument person, accountant, advisor and my best friend. Without her, I would not be the professional I am today. Cindy, thank you very much.

Other titles by this author:

http://www.cc4w.net/ebooks.html

Follow us on Facebook

Books available on amazon.com

SURVEYING MATHEMATICS MADE SIMPLE

MATH-SERIES TRAINING AND REFERENCE BOOKS / APPS

Printed - Digital - Apps
Many Titles to choose from.
www.cc4w.net